To creative Kids.

Because our future depends on them.

Robonauts® by NASA. RX78® by Sunrise Inc. PICO® by Sandia National Labs. Industrial Robot by Kuka® Robotics. HEXBUG® is a trademark of Innovation First Labs.Some pictures and information are property of their respective author, manufacturer company and/or copyright owner. They are cited on this book because of their relation and relevance. The information and brand names used here are solely for educational purposes and to preserve information about robot history for future generations.

Latin-Tech Inc
WWW.LT-AUTOMATION.COM

ISBN: 978-0-615935-64-5
PH: 305 848 3517 USA

Robot Story
Miami, FL, USA

Table of Contents

Famous Robots from Movies

The following Robots are famous because they have appeared in movies.

We have tried to give you the name of the first one (person or company) who developed the concept or name of the Robot.

Some characters are based on a book so, in this case the writer of the book is the creator of the Robot.

In some other cases, when the Robot is conceived in a company by an industrial designer, then the designer is considered the Robot´s creator.

Maria

Metropolis
UFA
Thea Von Harbou and Fritz Lang
1927

Maria is one of the first Robots in a movie.

She is one of the main **characters** in the German movie from 1927, Metropolis.

She is an android which **resembles** a girl named Maria.

This female Robot **encourages** the city workers to **rebel** against the high class owners of the company.

Maria is a Robot which looks like a human.

Nowadays she is used as a **reference** for other movies including Robots.

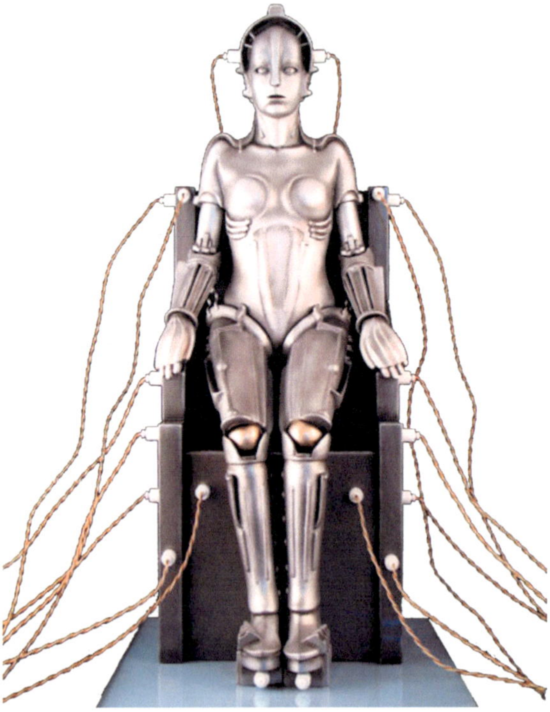

Gort

The Day the Earth Stood Still
20th Century Fox
Harry Bate
1951

Gort is an eight-feet tall Robot which came on a **flying saucer** to protect his friend Klaatu.

It has a powerful **beam**, coming out of his eyes, which uses when it wants to destroy something.

Its face has no mouth.

It **disintegrates** itself into tiny pieces which can destroy everything they touch.

Gunslinger

Gunslinger
Metro Goldwyn Mayer, Irving Allen and Block Adler
1973

Gunslinger is an android Robot which is very fast with guns.

It' s programmed to participate in **duels** that never wins.

It belongs to an amusement park where Robots serve all the visitors in whatever they want.

Because of a computer virus, all the Robots from the park began to have a wrong operation.

Gunslinger and other Robots kill some visitors at the **amusement** park.

Technicians and supervisors try to put the park under control by **shutting down** the power, but it doesn´t work.

Robbie

The Forbidden Planet
Metro Goldwyn Mayer
Robert Kinoshita
1959

Robbie was one of the first Robots to appear in movies.

It has no face but many sensors which help it to move around easily. Its 7 feet tall.

It appeared for the first time in a **science fiction** movie named "Forbidden Planet"

After this movie, he became famous and regulary appeared in other TV series, movies and shows.

The Robot was designed by Robert Kinoshita.

R2D2 and 3P0

Star Wars
20th Century Fox
George Lucas
1977

These two Robots became very famous after the successful movie Star Wars.

R2D2 (spelled Artoo-Detoo) is a **wheeled** Robot supported on three legs. It's loaded with many **tools**, arms and **sensors**.

C-3PO (spelled See-Threepio) is R2D2's friend. It´s a **protocol** android designed to serve humans. It´s able to speak in millions of **communication** forms.

Marvin

The Hitchhiker´s Guide to the Galaxy
BBC
Douglas Adams
1981-2005

Marvin is a **defective** Robot unit manufactured by
Sirius Cybernetic Corporation.

Marvin is an android with two different feelings: **Depression** and **Boredom**. This happens because his powerful brain is not fully used.

Every task appears to be so simple that Marvin becomes bored.
His brain is as big as a planet.

Even the biggest calculations become very easy for him. To **cheer** him up, some of his friends at the Heart of Gold Starship, try to keep him busy.

Ash

Alien
20th Century Fox, Dan O Bannon and Ronald Shusett
1979

Ash is a very sophisticated android. It´s a Science Officer, part of the crew aboard the **starship** Nostromo.

When a new life form is discovered in any far planet, it receives **secret** orders to take one **alien** to earth, no matter if the crew does not **survive**.

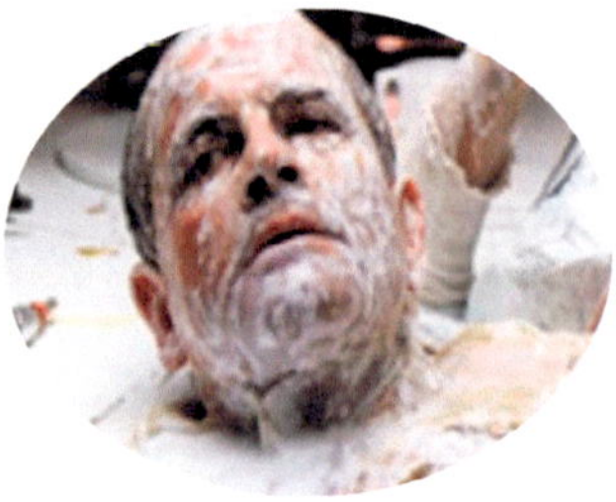

It **disobeyed** the captain's orders and put everybody in a dangerous situation.

Ash looks and behaves like a human.

Hector

Saturn 3
ITC Entertainment
John Barry
1980

Hector is an **experimental** Robot which arrives to one of the **moons** from the planet named Saturn.

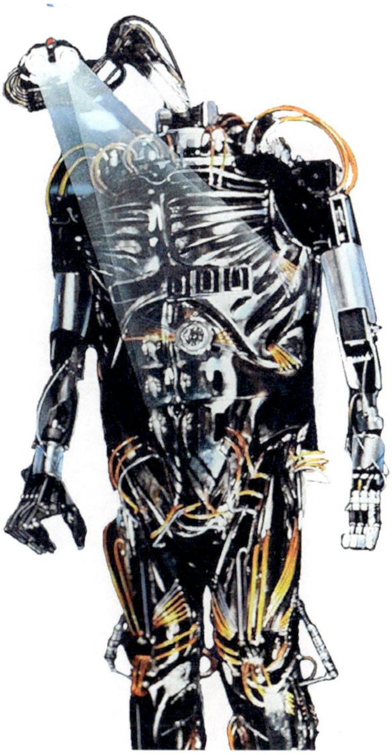

There, a **couple** of scientists have been working on growing food using **hydroponics** to feed the **overpopulated** earth.

Hector is a 7 feet tall Robot, assembled by a technician with **mental** problems named Benson.

The Robot has an artificial intelligence based on human brain **tissue**, which can be programmed using a **socket** on Benson´s neck.

He starts acting according to Benson´s behaviour.

Terminator

Terminator
Orion Pictures
James Cameron, Gale Anne Hurd, William Wisher, Jr
1984

Terminator is a Robot manufactured by a company named Skynet.

The Terminators are android Robots with a perfect human look.

They can smell and **sweat** to easily **infiltrate** among people.

Its main goal is to eliminate all type of **resistance** until they can **exterminate** to all the humans.

The first terminator was sent, from the future (2029) to the present times, to kill a girl named Sarah Connor.

Sarah will be the mother of the **leader** of the resistance in the future.

This page has been
left blank intentionally.

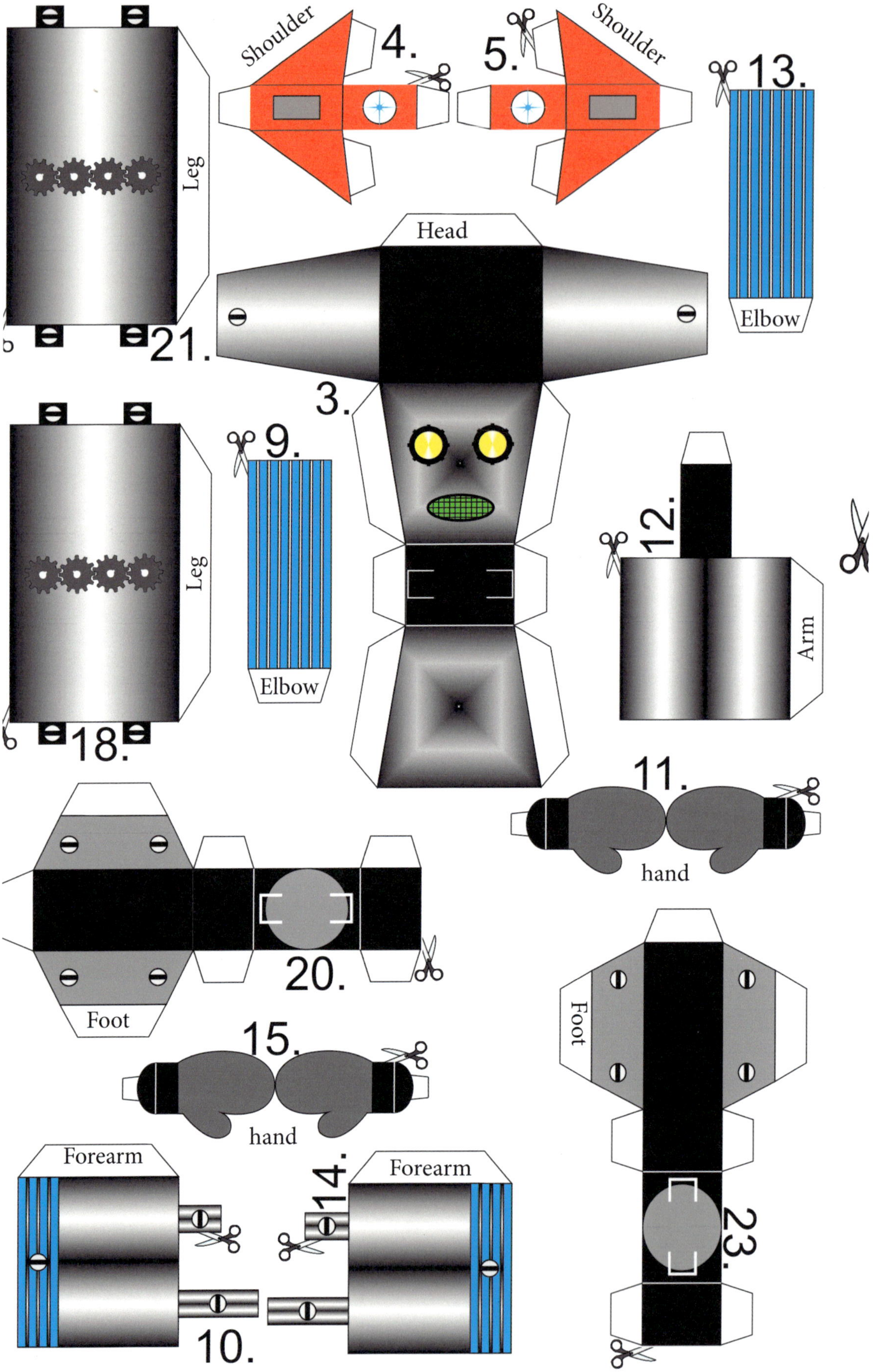

Shoulder
4.
5.
Shoulder
13.
Leg
21.
Head
Elbow
3.
9.
12.
Arm
Leg
Elbow
18.
11.
hand
20.
Foot
Foot
15.
hand
Forearm
14.
Forearm
23.
10.

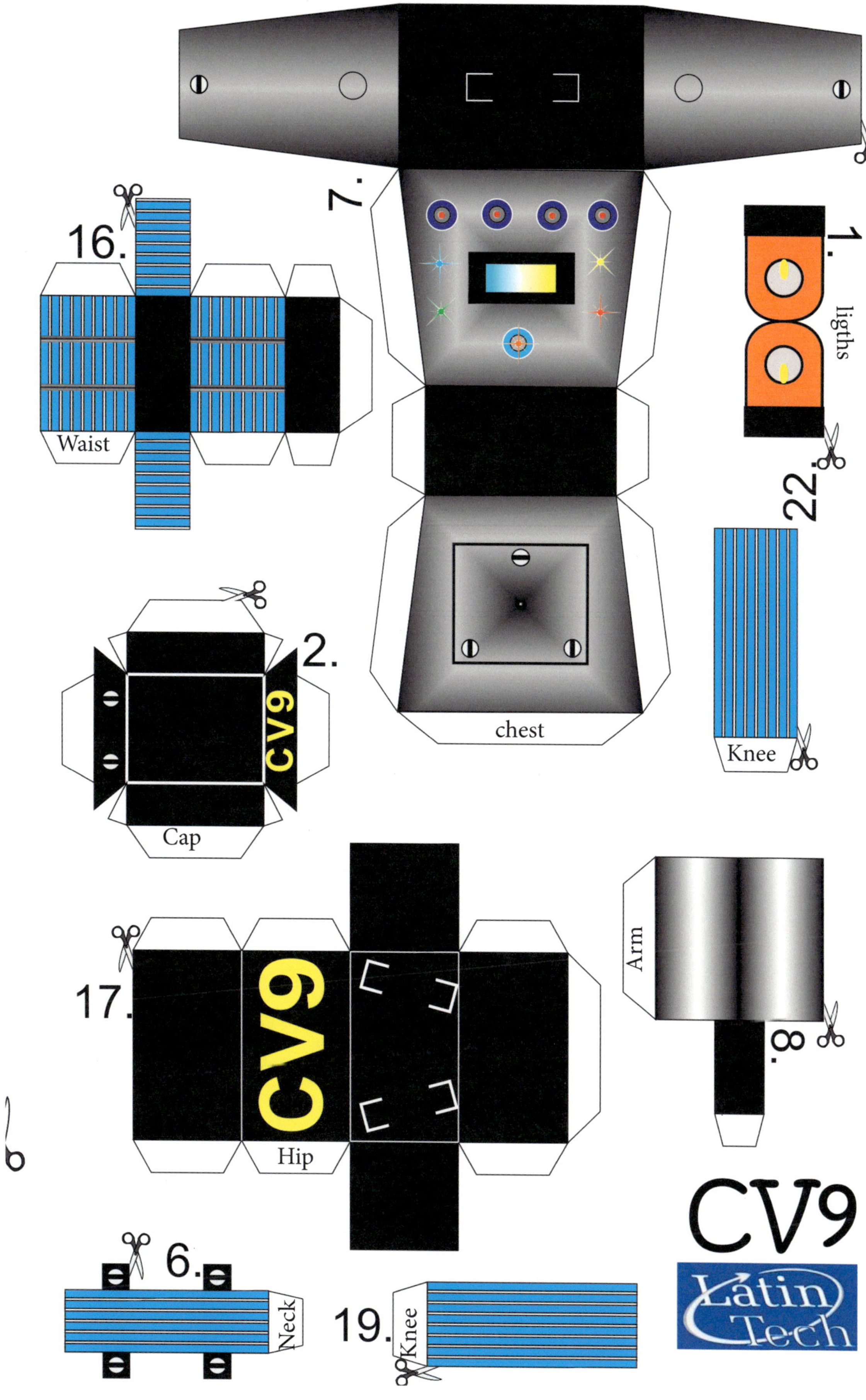

16.
Waist
7.
chest
1.
ligths
22.
Knee
2.
CV9
Cap
17.
CV9
Hip
Arm
8.
6.
Neck
19.
Knee
CV9
Latin Tech

This page has been
left blank intentionally.

Jhonny 5

Short Circuit
TriStar Pictures
S.S Wilson and Brent Maddock
1986

Number 5 is a **prototype** of a wheeled Robot, made by Nova Laboratories. This Robot is for **military** operations.

While it was charging, it receives a **power surge**.

This changes its program and produces a **malfunction** allowing him to develop feelings and a sense of protecting human lives.

BB

Deadly Friend
Warner Bros
Diana Henstel
1986

BB is a Robot with **Artificial** Intelligence which can only pronounce its name and, sometimes, act on its own.

It was created by Paul, a young and bright kid who moved to a new city to attend highly **specialized courses** on **Neurology** and Artificial Intelligence at a very **prestigious** science college.

BB is destroyed by a Paul's neighbor.

Samantha, one of Paul's friends, is about to die so he decided to use BB's Artificial Intelligence **chip** to save her life.

Samantha becomes a human body with a Robotic brain.

Robocop

Robocop
Orion Pictures
Edward Neumeier and Michael Miener
1987

Murphy is a policeman who has a beautiful family.

After being attacked by some **criminals**, his body is used by a company named Omni Consumer Corporation, as part of a new **development** in Robotics.

Since it is part human and part Robot it becomes a Cyborg with an **inner** conflict.

Its brain is programmed to act as a cop.

Andrew

Bicentennial Man
Pictures Touchstone/ Columbia pictures
Isaac Asimov
1999

Andrew is an android bought by the Martin family to help with the **housekeeping** activities.

They don't know that Andrew is special.

He develops emotions and shows **creativity**.

Within four **generations** later, he tries to become human.

The Iron Giant

Iron Giant
Warner Bros
Ted Hughes
1999

This Robot accidentally falls in a small town named Rockwell.

Hogwarth, a nine years old child finds it and later becomes his friend and teacher.

The Iron Giant is hunted by the Army. Hogwarth tries to avoid the conflict and explains the situation.

The Army launches a missile against the Robot. Hogwarth understands that the Robot must **sacrifice** itself to save the people in the town.

The Robot flies away and explodes in the air after being reached by the **missile**.

But then there is a surprise for Hogwarth's sadness...

David

Artificial Intelligence
Dream works/Warner Bros
Brian Aldiss
2001

David is part of a very intelligent **line** of Robot kids which can show emotions.

It was bought by a mother whose son has a very **rare** sickness that later becomes **curable**.

When the real son gets healthy, David is forced to **survive** alone in a world where Robots are not welcomed.

His intelligence and logic makes him to look for his **creator**.

He wants to recover his loving mother.

Sonny
I Robot
20Th Century Fox
Isaac Asimov
2004

"I, Robot" is a movie **inspired** by one of Isaac Asimov's books.

SONNY is an android Robot **accused** of killing his scientific creator and also breaking the Three Robotic Laws.

SONNY belongs to a new **generation** of Robots from the company US Robotics. He later becomes a **fugitive**.

Rodney Copperbottom

Robots
20th Century Fox
Chris Block and William Joyce
2005

Rodney is a Robot child, who always wanted to be an **inventor**.

One of his inventions generated a problem with his father's boss which forced him to leave his small town and go to Robot City.

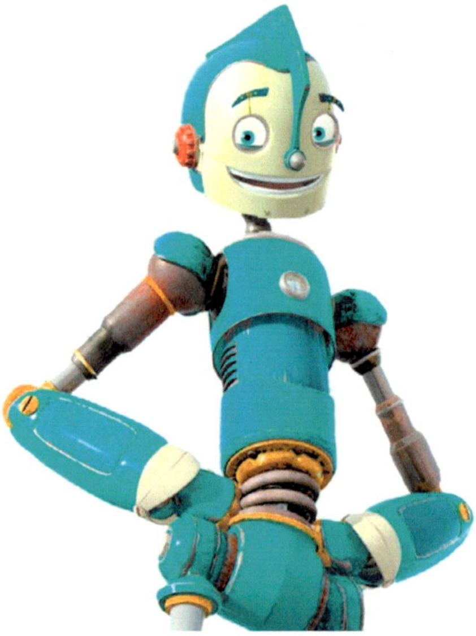

In this city lives a Robot, named BigWeld, who is admired by Rodney.

There, he makes new friends: the Rusties.

Wall-E

Wall-E

Pixar Animation Studios/ Walt Disney Pictures

Andrew Stanton and Pete Docter.

2008

A **Mega Corporation** made the people buy in large quantities.

People destroyed the planet by **covering** the entire world with trash.

For this reason, they had to leave the planet for five years.

Wall-E is a Trash **compactor** Robot left behind to **clean up** the Earth.

Atom

Real Steel
Dreamworks Pictures
2012

In the year 2020, only Robots are allowed to fight because it has been **forbidden** to humans.

When looking for Robotic pieces in a **junkyard**, a little boy named Max, discovers Atom the Robot.

In the past this Robot was used for **box training**.

Max helps his father, to train the Robot in order to participate in boxing competitions.

Atom is very strong and survives to several fights against many other very powerful Robots.

Vocabulary

Accused: Person charged by any wrong action.

Alien: Someone out of the earth.

Amusement: made to have fun or enjoy.

Artificial: Something made by humans.

Beam: A ray of light.

Boredom: State of feeling bored.

Box, Boxing: Sport where two opponents fight against each other.

Characters: Star or main actor in a film.

Cheer: Shout for joy or support.

Chip: Electronic integrated circuit.

Communication: Way to change information with others.

Compactor: Device that presses the thrash to have a smaller package.

Couple: Two people.

Clean up: restoring the order and putting everything neat.

Covering: A thing used to protect something.

Creativity: Having the ability to make original or imaginative things.

Creator: A person or thing that brings something into existence.

Criminal: Person who acts out of law.

Curable: Sickness or condition able to get cured.

Defective: Not working properly.

Depression: A feeling of being unhappy.

Development: Process of creating a new product.

Disintegrate: Breaking into several small pieces.

Disobey: Fail to obey.

Duels: Two person face each other with weapons.

Encourage: To suggest to people to do something.

Experimental: Something you don't know how it works or reacts.

Exterminate: To kill a group until there is no one left.

Flying saucer: Type of Unidentified Flying Object (UFO).

Forbidden: Something you are not allowed to do, somewhere your are not allowed to go.

Fugitive: That runs away from the law or a group.

Generations: The time that pass between the life of a person and the birth of that person´s Kids.

Housekeeping: Help and organization of home and all the works involved.

Hydroponics: Method to grow food without soil but with water.

Junkyard: A place where scrap or used parts are collected

Infiltrate: To become part of a group without anyone noticing.

Inner: Inside your mind.

Inspired: The origin of an idea.

Inventor: Person who invents things as a main job.

Vocabulary

Junkyard: Place where you just find trash.

Leader: Someone who directs a group.

Line: Similar products with minor differences.

Malfunction: Not working properly.

Mega Corporation: A super big company.

Mental: Relating to disorders of the mind.

Missile: A weapon with explosives thrown and directed to a target.

Military: That involves soldiers and armed forces.

Moon: A natural satellite of any planet.

Neurology: Science which studies diseases of the brain, spinal cord, nerves and muscles.

Overpopulated: Place with many people living in it.

Power surge: High power overcharge that generates malfunction.

Prestigious: Respected and admired.

Protocol: A way of correct conduct y behavior.

Prototype: The first Robot being made.

Rare: Strange, not common.

Rebel: To refuse to obey.

Reference: Something or someone to compare with.

Resemble: To look like someone.

Resistance: Opposition to something or someone.

Sacrifice: To give your life for something or someone.

Saucer: A small round flat dish.

Science Fiction: Story based on science and technological advances.

Secretly: Doing or telling soemthing without others knowing.

Sensors: Devices that help to detect what is around you.

Shutting Down: Removing the electricity.

Socket:A hole where you can connect an electrical equipment.

Specialized courses: Courses that are not common to find.

Surge: A sudden increase of electricity level.

Survive: To remain alive.

Sweat: Liquid that comes out of the skin.

Technicians: Persons expert in certain science.

Tissue: Different types of material of which animals or plants are made.

Tools: Devices you use to assemble or make things.

Training: Exercise that someone does regularly to practice.

Wheeled: Something that has wheels.

CV9

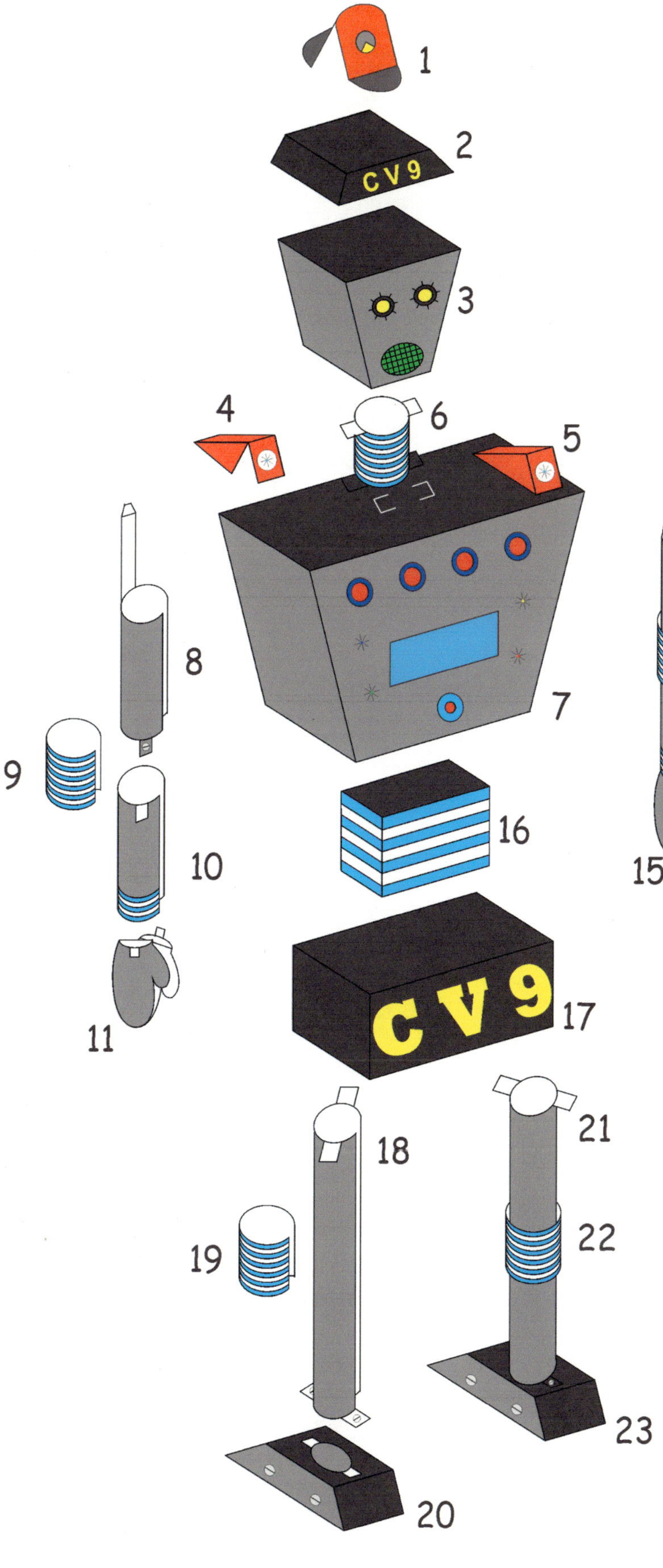

Made in the USA
Monee, IL
05 November 2026

67382718R00019